Capstone Short Biographies

Women in Engineering Careers

by Jetty Kahn

Consultant:
Jill Charpentier
Minnesota Section
Society of Women Engineers

CAPSTONE
HIGH/LOW BOOKS
an imprint of Capstone Press
Mankato, Minnesota

Capstone High/Low Books are published by Capstone Press
818 North Willow Street • Mankato, Minnesota 56001
http://www.capstone-press.com

Copyright © 1999 Capstone Press. All rights reserved.
No part of this book may be reproduced without written permission from the publisher. The publisher takes no responsibility for the use of any of the materials or methods described in this book, nor for the products thereof.
Printed in the United States of America.

Library of Congress Cataloging-in-Publication Data
Kahn, Jetty.
    Women in engineering careers/by Jetty Kahn.
      p. cm. — (Capstone short biographies)
    Includes bibliographical references and index.
    Summary: Presents an introduction to engineering followed by brief biographies of the following women engineers: Amy Alving, Cynthia Barnhart, Martha Gray, Jill Morgan, and Karen Zais.
    ISBN 0-7368-0013-1
    1. Women engineers—Biography—Juvenile literature. [1. Engineers. 2. Women—Biography.] I. Title. II. Series.
TA157.K23 1999
620´.0082´0922—dc21
[b]   98-1787
                                                                                     CIP
                                                                                      AC

Editorial Credits
Colleen Sexton, editor; Timothy Halldin, cover designer; Sheri Gosewisch,
    photo researcher

Photo Credits
Amy Alving, 10, 13, 14
Boeing, 34
Cynthia Barnhart, 16, 21
Images International/Erwin C. Nielsen, 4, 22
Jill Morgan, 32, 37
Karen Zais, 38, 41, 42
Martha Gray, 28
Photo Network/C. Ehlers, 27
Rainbow/Dan McCoy, 30
Shaffer Photography/James L. Shaffer, 8
Ted Fitzgerald, 24
Tom Foley, cover
Unicorn Stock Photos/Scott Liles, 7; Aneal Vohra, 19

Table of Contents

Chapter 1 Engineering 5
Chapter 2 Amy Alving 11
Chapter 3 Cynthia Barnhart 17
Chapter 4 Martha Gray 25
Chapter 5 Jill Morgan 33
Chapter 6 Karen Zais 39

Words to Know.. 44
To Learn More 45
Useful Addresses 46
Internet Sites 47
Index 48

Chapter 1
Engineering

Engineers use science and math to design useful structures and to solve problems. Engineers make plans for machines, bridges, and medical tools. They create new processes. They also develop testing methods to improve the way machines and tools work.

Engineering Fields

Engineering has several main fields of study. Mechanical engineers work with machines and engines. Chemical engineers find ways to process and handle liquids and gases. Civil engineers plan city structures such as roads, bridges, buildings, and railroads. Electrical

Civil engineers plan city structures such as bridges.

engineers design electrical systems, computer systems, and communication systems. Industrial engineers organize machines, people, and materials in factories. They find the best ways for factories to make products.

Some engineers work in specialized fields. Aerospace engineers design aircraft and find new ways to help aircraft fly faster and safer. Biomedical engineers work closely with doctors and biologists to design medical products. They also study medical problems and create machines or processes to solve the problems. Agricultural engineers design farms and food processing equipment. They design systems that bring water to fields and remove extra water from fields.

Education
People who like to take things apart and put them back together make good engineers. Engineers are curious about how structures are built and how machines work. People combine this curiosity with an education in science and math to become engineers.

Aerospace engineers find ways to help aircraft fly faster and safer.

People who want to be engineers must attend college. Students take general science, math, and engineering classes during the first two years of college. They then choose a particular engineering field to study. Engineers with four-year college degrees work mainly for businesses and the government.

Many engineers earn higher degrees such as master's degrees or doctorates. A doctorate is the highest degree given by a university. Engineers with these degrees often teach or conduct research. Researchers perform close and careful study of a subject. People often perform experiments when conducting research. Some engineers with higher degrees do research in very specialized fields of study.

Agricultural engineers plan farms.

Chapter 2
Amy Alving

Amy Alving sees the game of golf through the eyes of an engineer. She knows that the dented surface of a golf ball helps carry it through the air. A ball with a smooth surface would travel only half the distance of the dented ball. Alving's background in aerospace engineering helps her know how golf balls and other objects fly through the air.

    Alving earned a bachelor's degree in mechanical engineering from Stanford University in Palo Alto, California. She also earned a doctorate in mechanical and aerospace engineering from Princeton

Amy Alving is an aerospace engineer.

University in Princeton, New Jersey. Today, Alving teaches aerospace engineering at the University of Minnesota in Minneapolis.

Air Resistance
Aerospace engineers figure out ways for aircraft to travel through the air with less resistance. Air resistance is the force of air pushing against objects. A typical aircraft needs a lot of fuel to move through air resistance. Large amounts of fuel can be saved if air resistance is decreased.

   Alving designs airplane wings that decrease air resistance. The shape of a wing affects how it moves through air. Alving tests different wing shapes in a wind tunnel. Air blows around the wings in the tunnel. Computers help Alving use measurements from the tests to figure out which wing shape is best.

Turbulence
Aerospace engineers think of air as a liquid. Like liquid, air can move and change shape. Liquid sometimes starts to swirl when it

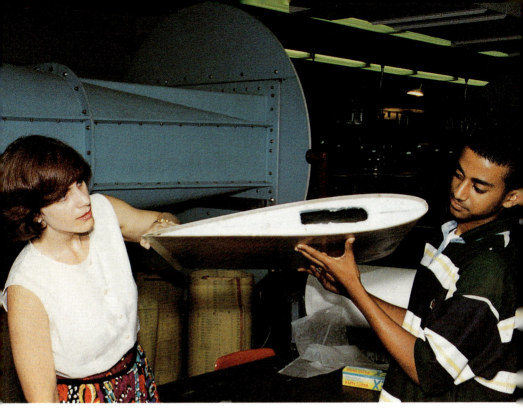

Amy Alving looks at the shape of an airplane wing.

moves quickly. This swirling is called turbulence. Air flowing around an aircraft also swirls and creates turbulence. Uncontrolled turbulence can quickly slow down an aircraft.

Alving studies turbulence. She looks for ways to break up turbulence so that aircraft can move smoothly through the air. Alving experiments with a special ribbed plastic. The plastic has a wavy surface. Alving covers

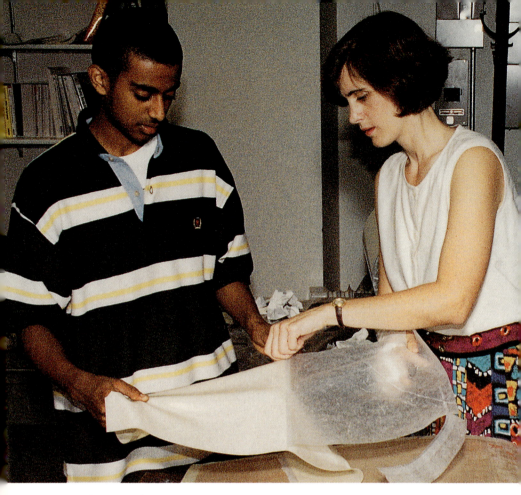

Amy Alving is developing a new ribbed plastic. The plastic helps airplane wings move more easily through air.

aircraft wings with the ribbed plastic. The plastic helps wings break up turbulence and move more easily through air.

Powerful computers help Alving and other scientists study turbulence. The computers use measurements gathered in experiments to quickly solve hard math problems. The results help Alving understand turbulence. Her findings may lead to planes, cars, and even bicycles that are less air resistant.

Science and Sailing
Alving also studies the way sailboats move through water and air. Scientists noticed that sharks and other fish have ribbed skin. The skin reduces water resistance to help them swim quickly through water. Alving used this idea to test a ribbed plastic coating on the outside of boats.

One of the first sailboats to have the ribbed plastic was in a race called the America's Cup. The ribbed plastic helped the boat cut quickly through water. The boat won 11 out of 12 races. Officials banned the use of the ribbed plastic from sailing races. They thought it gave some boats an unfair advantage.

Alving is designing a special ribbed sail. The ribbing may make the sail more effective and help sailors win races.

Chapter 3
Cynthia Barnhart

Five-year-old Olivia made a large birthday card for her grandmother. Olivia and her mother brought the card to United Parcel Service (UPS). UPS delivers letters and packages to people around the world. Olivia's mother is Cynthia Barnhart. She helped UPS find ways to better organize the shipment of packages like Olivia's.

Barnhart is a professor at the Massachusetts Institute of Technology (MIT) in Cambridge, Massachusetts. Barnhart conducted a research project at a UPS center. She used her engineering skills to improve the company's process for shipping packages.

Cynthia Barnhart is a professor at the Massachusetts Institute of Technology.

Barnhart's education prepared her for teaching and research. Barnhart earned a doctorate in civil engineering and transportation from MIT. Barnhart applies her engineering knowledge to the field of transportation.

Creating Transportation Systems

Barnhart helped UPS set up effective transportation systems. She tried to find the best ways to move goods from one place to another in a certain amount of time. Transportation systems use airplanes, trains, trucks, and ships. Well-organized transportation systems help UPS and its customers save money on mailing costs.

Barnhart considers many elements when she creates transportation systems. She knows that a problem or an unexpected situation can affect the entire system. Barnhart looks at entire transportation systems. She finds ways to make systems organized and safe.

Barnhart creates transportation systems on computers. She sets up computer programs that figure out how to move packages from one place to another. Each package needs to arrive

Highways are part of transportation systems.

in a certain amount of time. Customers choose how fast they want their packages to be delivered. Customers pay different prices for different delivery times.

Computer programs keep track of the trucks or airplanes that will carry a package. These programs also figure out how many other packages could be delivered on the same trip. UPS saves money when many packages can be grouped together for delivery.

Barnhart's computer programs also create backup plans. Backup plans allow UPS to deliver packages on time when unexpected events happen. For example, a heavy snowstorm might block roads so that trucks cannot get through. A backup plan may say to put packages on an airplane. The computer program also helps reschedule flights if airplanes cannot take off on time. Backup plans help save time and money.

Barnhart also considers safety when she creates transportation systems. She looks at the number of planes using certain flight paths. Too many planes in one area could cause accidents. She considers the number of hours that flight crews and truck drivers can safely work. These workers may get tired and make mistakes if they work too many hours.

Cynthia Barnhart created a computer program for UPS to organize the shipment of packages.

Teaching and Travel

Barnhart teaches students at MIT about transportation systems. She conducts many transportation studies as part of her job there. She writes reports about the results of her studies to share with other engineers.

Barnhart also travels to other countries as part of her work. She helps these countries solve their transportation problems.

Barnhart worked with scientists in the Central American country of Panama during one trip. She helped the engineers plan a way for many ships to move through the Panama Canal quickly. The Panama Canal allows ships to travel between the Atlantic and Pacific Oceans.

Cynthia Barnhart helped engineers in Panama route ships through the Panama Canal.

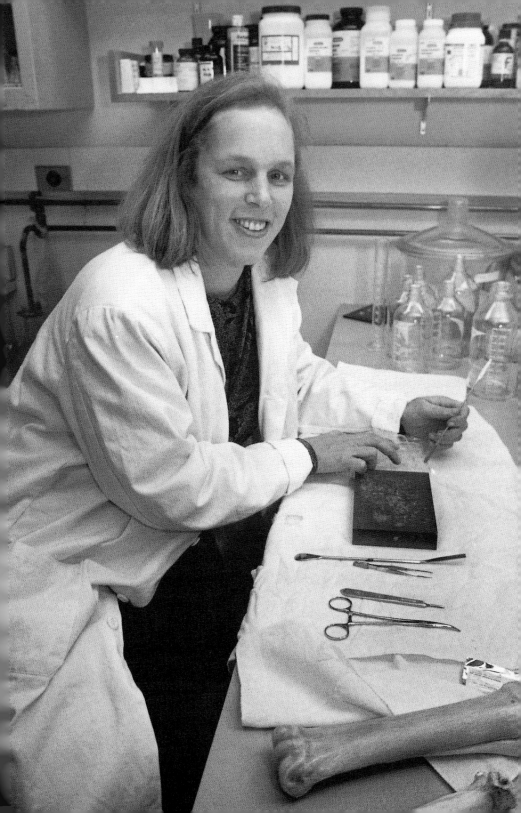

Chapter 4
Martha Gray

Martha Gray sees the human body through the eyes of an engineer. The body has matter and energy that work together to help humans move. Gray uses her engineering knowledge to solve medical problems.

Gray teaches electrical and medical engineering at MIT. To prepare for this career, Gray earned a bachelor's degree in computer science from Michigan State University. She also has a doctorate in medical engineering from the Harvard-MIT Division of Health Sciences and Technology in Cambridge, Massachusetts.

Martha Gray studies the way matter and energy work together to help the human body move.

Cartilage Damage

Gray is an expert in cartilage damage. Cartilage is strong tissue that connects bones in people and animals. Damaged cartilage is torn, chipped, or worn down.

Gray helps people who have damaged cartilage in their knees. The knee joint is the point at which the two large leg bones meet. A thin layer of cartilage acts as padding between the bones.

Extra pads of cartilage in the knee keep it steady. The additional cartilage also protects the knees. Knees carry much of people's weight when they stand or walk. Pressure on the knees increases when people run or jump.

Torn cartilage is a common sports injury. Repeated side-to-side movements in sports like tennis or basketball can rip the cartilage pads. The pads also can wear down when people run on hard surfaces.

Cartilage damage causes pain and swelling in the knees. People with this kind of damage often cannot move their knees well. Cartilage damage can lead to arthritis. Arthritis is a

People who play basketball may damage the cartilage in their knees.

disease that causes joints to swell. Arthritis is painful and hard to treat.

Cartilage Studies

Gray works with a team of other scientists to study cartilage. Gray studies how well cartilage cells withstand different amounts of pressure. She looks for ways to prevent cartilage damage and stop the onset of arthritis.

Gray studies ways to find minor damage in cartilage early. This could help prevent major damage later. A magnetic resonance imaging (MRI) machine has helped her studies in this area. MRI machines take pictures of the inside of the body. The pictures show soft tissues such as muscles and cartilage.

Gray found a way for MRI machines to measure the amount of energy in cartilage. The energy of healthy cartilage creates bright MRI pictures. Damaged cells have less energy and create dull MRI pictures. Gray uses MRI machines to see whether cartilage damage exists or if damaged cartilage is healing.

Martha Gray studies ways to find minor cartilage damage early.

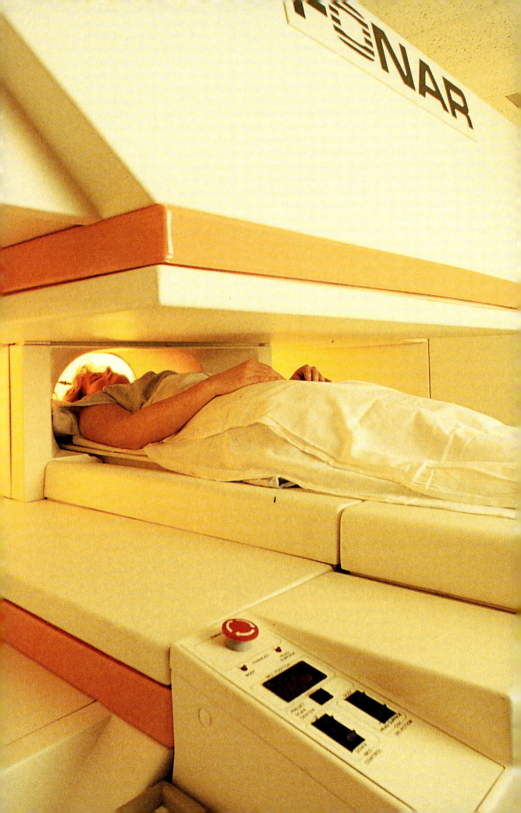

Gray's research helps people stop damage to their knees. People can find out earlier if they need to change the way they exercise. They may do sports such as swimming or biking. These activities do not put as much pressure on knees. People may decide to run on a softer surface such as grass instead of on hard roads.

The use of MRI machines helps doctors learn how to better treat patients with knee damage. Pictures from the MRI machine show if certain exercises are healing or harming cartilage. The tests also show doctors how patients are recovering from knee operations.

Martha Gray uses MRI machines to look for cartilage damage in people's knees.

Chapter 5
Jill Morgan

Jill Morgan studied languages in college. She wanted to be a teacher or a translator. A translator changes words from one language to another. But Morgan's career plans changed when she started talking to some engineers.

One day she took her stepfather to a meeting at a university. Her stepfather was in a wheelchair. The wheelchair broke and two engineering professors rushed to help. Morgan asked them about their work. She was interested in what they did.

Morgan decided to learn more about engineering. She began taking classes in math and science. She earned two bachelor's

Jill Morgan is a mechanical engineer.

degrees from the University of Minnesota. One was in languages and the other was in mechanical engineering.

Testing Airplanes

Today, Morgan works for MTS Systems Corporation. This company makes machines to test many kinds of materials and products. The machines test metals, concrete, cars, airplanes, bridges, and even theme park rides. The results show if the products are safe and effective.

Morgan is a senior project engineer in the aerospace group at MTS. Her customers are companies that build airplanes. She oversees the manufacture and delivery of testing machines for her group. Morgan makes sure the machines will perform the tests well.

Airplane builders use the testing machines to help build better planes. The machines provide the same conditions that airplanes experience in the air. Airplanes must endure great temperature changes as they move from the ground to the sky. The wings and body of the plane must withstand shaking during air

Jill Morgan helps develop testing machines for large airplane manufacturers.

turbulence. Airplanes also must withstand air pressure changes during takeoff, flight, and landing. The machines that Morgan's company supplies help builders see how planes react to these conditions. The results help airplane manufacturers make planes stronger and safer.

Engineering and Languages

MTS has offices all over the world. Morgan uses her foreign language skills to help MTS communicate with customers in other countries. But she chooses not to travel to these countries because she has young children.

Morgan translates documents written in French or Spanish. This work and her knowledge of languages has helped advance her career in engineering.

Jill Morgan chooses not to travel in her job because she has young children.

Chapter 6
Karen Zais

Karen Zais likes to bake. She stirs together ingredients to make treats for her family and friends. But at work she helps make treats for people all over the world. Zais works for The Pillsbury Company in Minneapolis, Minnesota. Pillsbury is a manufacturing company that prepares baked goods and many other foods.

Education and Training
Zais gained a variety of experience before working at Pillsbury. She has a bachelor's degree in mechanical engineering from Michigan Technological University in Houghton, Michigan. She also had summer

Karen Zais uses her mechanical engineering background to help prepare food products.

jobs during college that helped her prepare for a career in engineering.

One summer she worked for a defense company to study safety on submarines. Another summer she worked for a car-part manufacturer. She studied instruments that make air bags fill up in cars during crashes. Zais also spent a summer designing work stations for employees of a manufacturing company. The work stations prevented employees from hurting their backs, necks, and wrists.

Zais worked for an oil-refining company after graduating from college. She directed a team that built towers and set up pumps. She often was in charge of a 20-person crew.

Food Engineering
At Pillsbury, Zais uses computers to make sure machines that manufacture food work correctly. The machines mix, package, and freeze food. Zais and other engineers use the computers to make sure the food is prepared right.

The machines usually work correctly. But sometimes the engineers find problems.

Karen Zais uses computers to make sure machines mix, package, and freeze food correctly.

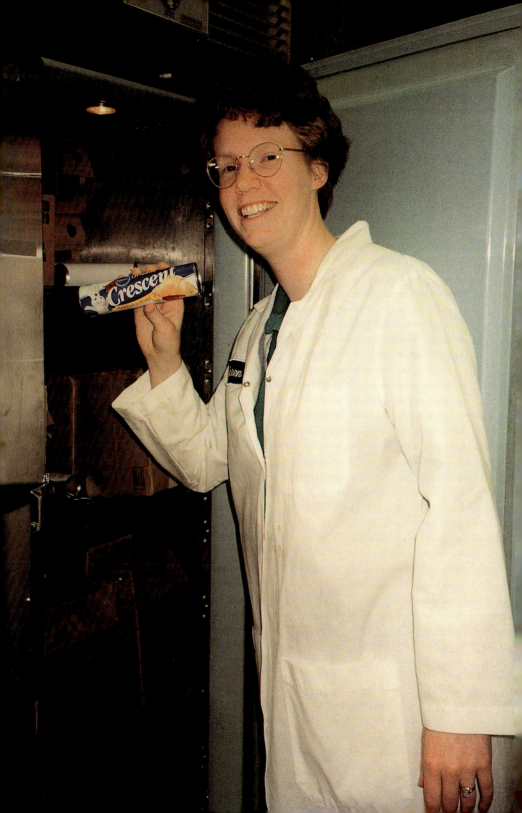

The giant mixers sometimes beat all the air out of batter. Cakes and muffins turn out too thick when this happens. Zais and a team of engineers and food scientists work to solve the problems. They work to correct the machines.

Zais also pays close attention to factory temperatures. Some ingredients can be damaged or destroyed if the temperature is too high or too low.

The ingredients sometimes can create problems. For example, wheat flour is a main ingredient in most baked goods. One year, bad weather changed the quality of wheat flour. Too little protein in the flour made the baked goods heavy. Zais and other engineers changed the batter and the machines to correct the problem.

Karen Zais must find ways to correct machines when problems arise during processing.

Words to Know

air resistance (AYR ree-ZIHS-tuhns)—the force of air pushing against objects

arthritis (ar-THRY-tuhs)—a disease that causes joints to swell

cartilage (CAR-tuh-lihj)—strong tissue that connects bones in people and animals

design (dee-ZYNE)—to make a plan of something that could be built

research (REE-surch)—close and careful study of a subject

translator (TRANZ-lay-tuhr)—a person who changes words from one language to another

turbulence (TUR-byoo-luns)—swirling winds that create strong air resistance; turbulence can quickly slow down an aircraft.

To Learn More

Evans, David, and Claudette Williams. *Building Things.* Let's Explore Science. New York: Dorling Kindersley, 1993.

Garratt, James. *Design and Technology.* New York: Cambridge University Press, 1996.

Goodwin, Peter H. *More Engineering Projects for Young Scientists.* Projects for Young Scientists. New York: Franklin Watts, 1994.

Morgan, Sally and Adrian Morgan. *Structures.* Designs in Science. New York: Facts On File, 1993.

Wilkinson, Philip. *Super Structures.* Inside DK Guides. New York: DK Publishing, 1996.

Wood, Robert W. *Science for Kids: 39 Easy Engineering Experiments.* Blue Ridge Summit, Penn.: Tab Books, 1992.

Useful Addresses

American Society for Engineering Education
1818 N Street NW
Suite 600
Washington, DC 20036-2479

Society for Canadian Women in Science and Technology
417-535 Hornby Street
Vancouver, BC V6C 2E8
Canada

Society of Women Engineers
120 Wall Street
11th Floor
New York, NY 10005-3902

Internet Sites

Aero Design Team Online
http://quest.arc.nasa.gov/aero/team/fjournals/
   zuniga/index.html

Conditions of the Knee
http://www.drmendbone.com/knee.htm

Engineering: Your Future
http://www.asee.org/precollege/default.htm

Society of Women Engineers
http://www.swe.org

Transportation Wonderland
http://education.dot.gov/k5/gamk5.htm

Index

aerospace engineering, 6, 11, 12
aircraft, 6, 12, 13
airplane, 12, 18, 19, 20, 35
arthritis, 26, 29

baked goods, 39, 43
boat, 15

cartilage, 26, 29, 31
chemical engineers, 5
computer, 6, 12, 14, 18, 19, 20, 25, 40

degree, 9, 11, 25, 39

experiment, 9, 13, 15

golf, 11

ingredients, 39, 43

math, 5, 6, 9, 15, 33
mechanical engineering, 5, 11, 35, 39
methods, 5
mix, 40
MRI machine, 29, 31

package, 17, 19, 20, 40
Panama Canal, 23
process, 5, 6

research, 9, 17, 18

structure, 5, 6
submarine, 40

translator, 33
transportation, 18, 20, 23
turbulence, 13, 15, 36

United Parcel Service (UPS), 17, 18, 20